AF394418

Business Skills Training
for Rural Sanitation
Entrepreneurs

TRAINER'S GUIDE

JOSHUA KIBET | DIANA MULATYA

Disclaimer

The Business Skills Training Manual for Rural Sanitation Entrepreneurs was made possible by the support of the American People through the United States Agency for International Development (USAID) under the terms of Contract No. AID-615-TO-15-00001. The contents of this training manual are the sole responsibility of the authors and do not necessarily reflect the views of USAID or the United States Government.

Use of the content for noncommercial use is authorized, provided the source is acknowledged.

ABBREVIATIONS AND ACRONYMS

KES	Kenya Shillings
KIWASH	Kenya Integrated Water and Sanitation Hygiene Project
SACCO	Savings and Credit Cooperative
SAFI	A concrete toilet with pre-cast rings for sub-structure
SATO	SAFI Toilet, a brand of LIXIL Corporation
USAID	United States Aid for International Development
VSLA	Village Savings and Loans Association

CONTENTS

Authors

Joshua Kibet is an Infrastructure Project Finance Expert and a Certified Small Business Development Advisor. He holds a Masters Degree in Financial Economics. For 17 years, Joshua has specialized in providing business advisory to community-based enterprises and technical assistance to infrastructure programs and enterprises in sectors such as water, sanitation, agribusiness, renewable energy and micro-finance. He has previously worked as a Consultant for the World Banks' Water and Sanitation Program (WSP-Africa) and Global Program on Output Based Aid, the EU Intra-ACP GCCA+ Climate Support Facility, and the Kenya Government to support flow of private finance to water and sanitation sector.

email: shuajok@gmail.com   Cell : +254 722 378 667   Twitter: @joshuakibet1

Diana Mulatya is a public health expert with over 10 years in design and development of water, sanitation and hygiene programs. She holds two Masters Degrees, in Environmental Management and Public Health in health and environment from the London School of Public Health. Her work has mainly focused on bridging research with practice, providing advisory services and developing evidence-based knowledge products to guide positive behavioral shifts in rural communities. She has co-authored peer-reviewed publications on diarrhea, upper respiratory tract infections and other diseases. Diana has served in several organizations including UNICEF, SNV - Netherlands as well as with the Kenyan government, providing technical assistance in delivery of water, sanitation and hygiene services.

email: dianamulatya@gmail.com   Cell: +254 720 322653

Contributors

First and foremost, we would like to thank our two reviewers: Mr. Shiva Narain Singh of UNICEF, Uganda and Ms. Emily Mutai, a communications and knowledge management expert for their constructive criticism and support in content creation.

This document would not be complete without the valuable contributions from our sanitation specialist — Harun Kariuki, Josephine Githinji, Juliet Akinyi, Linda Abong, Elsie Too Maitabel Okumu, Nancy Khatenje, Nancy Nyagoha, Roselyne Okwiri, Titus Kioko and Victorine Ndinda — who provided substantial input to case documentation and tested materials by training small-scale sanitation entrepreneurs in eight counties in Kenya.

Editing, Layout & Illustration

Cari Longman   Business Writing Consultant

Samuel Muigai   Illustrator

Rosalia Mumo   Graphics & Layout Designer

Introduction

Business Skills Training for
Rural Sanitation Entrepreneurs

Introduction

This manual was developed as part of a training package to support business development skills trainings for local sanitation entrepreneurs in Kenya. Financial and technical support was provided by the United States Agency for International Development (USAID) under the Kenya Integrated Water and Sanitation (KIWASH) project. KIWASH was a five year (2015–2020) project implemented by the Development Alternatives Incorporation (DAI) across nine counties. One of the key goals of KIWASH was to help trigger and activate demand for low-cost affordable sanitation technologies in rural and low-income communities.

1.1 The SAFI Latrines and SATO Technology

In 2016, a market survey conducted by KIWASH identified a concrete toilet with pre-cast rings for sub-structure (SAFI) latrines and toilet, a brand of LIXIL Corporation (SATO) technology as the most preferred and easily scalable sanitation products available in Kenya's rural market. These among other programmatic considerations such as, ability and ease of production, affordability or sustainability — informed the projects decision to promote these two technologies and trigger an increase in adoption by rural households.

By definition, SATO is a simple sanitation technology that can be retrofitted on a pit latrine to improve quality and aesthetics of the structure. It has several advantages to users; can be installed by local artisans, eliminates odor, smell and is easy to clean. The product uses minimal water for flushing, less than 500 ml per use. On the other hand, SAFI latrine is a low cost ventilated improved pit latrine constructed with reinforced circular pits which prevent the common problem of collapsing pits. Collapsing pits are widely experienced in areas with poor soil formation.

1.2 Objectives of the Manual

The overall objective of this manual is to equip sanitation specialists and public resource persons with the basic concepts and tools, to facilitate entrepreneurship and financial literacy trainings for start-up sanitation entrepreneurs in rural communities.

Specifically, this manual is designed to help participants:

- Learn the basic concepts of entrepreneurship and characteristics of successful entrepreneurs;
- Learn and practice essential marketing techniques for sanitation products and services;
- Develop money management competencies necessary to succeed as a small-scale entrepreneur;
- Build necessary leadership and management skills to grow successful sanitation enterprises.

1.3   For Whom Is This Manual Designed?

This manual is designed to strengthen, improve and effectively facilitate entrepreneurial knowledge and skills to set up, manage and grow a sanitation enterprise. These entrepreneurs are often community health volunteers and artisans who work within a Community Health Unit to promote sanitation and hygiene messaging and are keen to grow their own sanitation retail enterprises within their communities.

The manual may also be used by other facilitators implementing similar sanitation (marketing) programs in rural communities.

1.4 The Manual Development Process

This manual was developed over an eight-month period as part of the USAID's Kenya Integrated Water, Sanitation and Hygiene (KIWASH) project work in helping communities climb up the sanitation ladder using the sanitation marketing approach. The content was developed based on a training needs assessment carried out among 791 prospective sanitation entrepreneurs interested in starting SATO toilet retail enterprises in eight program counties (Figure 1).

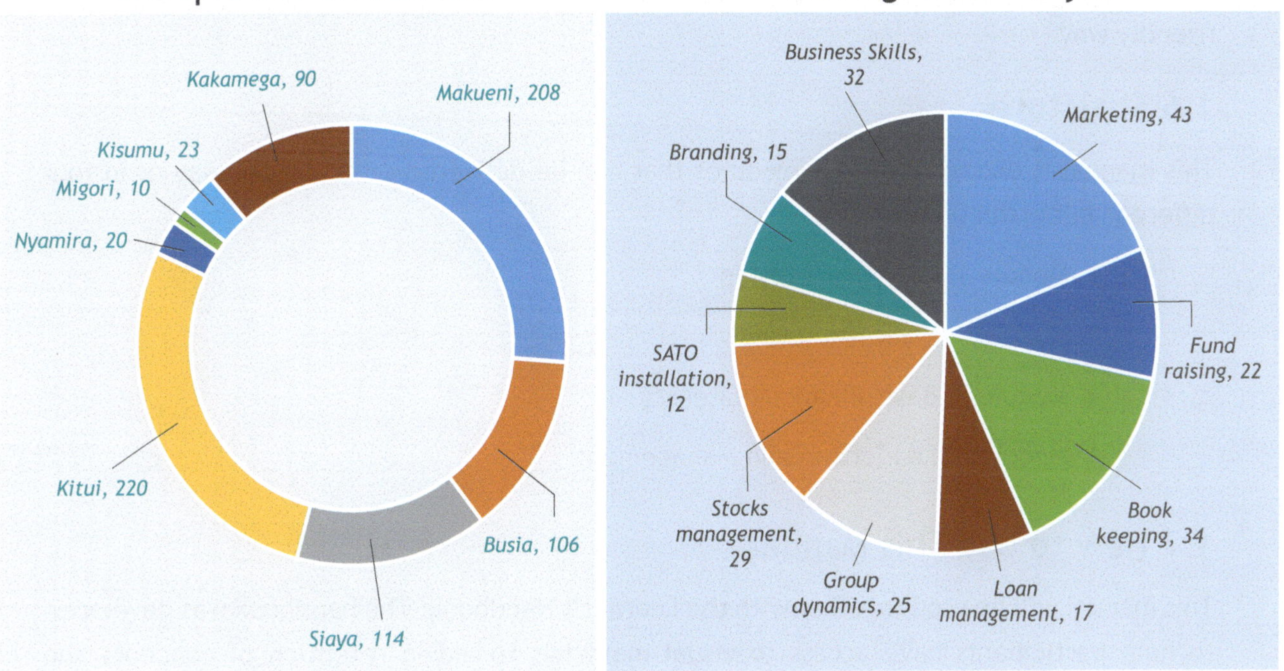

Figure 1: Sanitation Enterprise Training Needs Analysis Results, July 2019.

The curriculum was developed with substantial input from program specialists from the eight counties. The manual outline and draft content were reviewed during a one-day workshop in Kisumu, where the delivery process, trainer resources and methodologies were agreed upon.

1.5 Delivery Method

Conventional learning with a teacher or trainer lecturing in front of a group has proven to be less effective than learning approaches that actively engage participants in the learning process. The content in this manual focuses on a participatory learning approach, including training methodologies that enable effective facilitation, participatory and experiential learning for long-term knowledge retention.

Training participants are encouraged to take an active, engaged approach to learning. Regardless of the level of formal education, each participant has a valuable contribution to make, if encouraged to be an active partner in the learning process.

Based on this approach, the trainer acts as and is referred to as a "facilitator." They facilitate participatory and experiential learning, i.e., create the conditions and the environment in which trainees themselves will discover and practice new knowledge and skills.

By means of exercises, trainees will explore and share their own experience and knowledge, critically conceptualize and reflect upon solutions from which they will derive learning and ways for practical implementation of an enterprise.

Appropriate activities and images included in the manual will support learning in a student-friendly way.

1.6 Structure

This manual is divided into four modules that can be delivered either in one day or in four different in-person sessions:

- ♦ Module 1: Entrepreneurship
- ♦ Module 2: Basic Marketing
- ♦ Module 3: Basic Financial Literacy
- ♦ Module 4: Leadership and Management

1.7 How to Use This Manual

This manual will be used together with the Learner's Handbook. The handbook was developed to help participants have access to visual materials to aid in retention of concepts and practice using various exercises.

The manual is not exhaustive on entrepreneurship content. Trainers are encouraged to read widely on the subject, but also familiarize themselves with the organization and content of the Learner's Handbook before any training session.

Module 1

ENTREPRENEURSHIP

Business Skills Training for
**Rural Sanitation
Entrepreneurs**

Session Plan

Session Title	Entrepreneurship
Time	30 minutes
Objectives	At the end of this session, participants should be able to:
	◊ Define key concepts: *Entrepreneur, Entrepreneurship, Enterprise* and *Sanitation Enterprise*
	◊ Identify key characteristics and competencies of an entrepreneur
	◊ Define resources needed to start a business
Contents	◊ Defining an entrepreneur and entrepreneurship
	◊ Characteristics and qualities of an entrepreneur
	◊ Requirements for successful business start-up
Resources	◊ Picture Poster: Artisan Timothy Kulavi
	◊ Participants workbook
Methods	◊ Case study
	◊ Question and answer discussion
	◊ Experience sharing

Session Guide

Step 1: 10 minutes

The Concepts of "Enterprise" and "Entrepreneurship"

Training materials: Handout 1.1:
Concept Definitions

1.1 Start the session with an energizer (i.e. song, dance or metaphor). Introduce the session objective and the importance or benefits of running an enterprise. Ask participants if they know of any successful entrepreneurs in their community. What could be some of their motivation for being in business? 

- ♦ Earn extra income
- ♦ Create jobs (or alternative livelihood) for your family
- ♦ Utilize your skills
- ♦ Solve a problem
- ♦ Provide a demanded service/product

1.2 Inform participants that the term "enterprise" also means "business." An entrepreneur refers to a businessman/woman or a group engaged in a business activity. This could include a Community Health Volunteers (CHV's) retailing SATOs, an artisan constructing toilets, or a group buying and selling toilet products, etc.

1.3 Summarize the session by informing participants that starting an enterprise involves taking risk, with a motive to make PROFIT. This defines an entrepreneur. 

Step 2: 15 minutes

Characteristics of an Entrepreneur

Training materials: Handout 1.2
Case Study: Timothy Kulavi and Key Attributes of Entrepreneurs

2.1 Guide participants through the case of TIMOTHY KULAVI, a successful sanitation entrepreneur in Kakamega. (This may also be substituted for a local artisan entrepreneur success story.)

- ♦ Read out loud Kulavi's story. Ask participants to identify key attributes/ characteristics that made Kulavi a successful entrepreneur.
- ♦ Look over entrepreneurship attributes listed on Handout 1.2.

2.2   Ask some participants to mention a few of their personal "entrepreneur" characteristics. Begin a debate or discussion on which attributes are most important to running a successful sanitation enterprise.

2.3 Summarize this session with key attributes of Timothy:

- ♦ Focused
- ♦ Identifies business opportunities
- ♦ Patient
- ♦ Takes calculated risks
- ♦ Draws satisfaction by utilizing his own skills
- ♦ Passionate about sanitation

Step 3: 20 minutes

Things to Consider When Starting a Sanitation Enterprise
Training materials: Handout 1.3
Important Elements When Starting an Enterprise

3.1 Why do people start enterprises and fail? Discuss the reasons but with emphasis on the following elements that are important in starting an enterprise.

- ♦ Market opportunity
- ♦ Plan
- ♦ Capital
- ♦ Cashflows
- ♦ Motivation
- ♦ Profit

3.2   Ask some participants to brainstorm and share about the main types of businesses operating as sanitation enterprise in their area and how they started. Highlight the range of products or services offered;

a. Producers of prefabricated latrine products
b. Retailers of manufactured latrine products, e.g., SATO products, plastic slabs
c. Construction material shops/stockiest, e.g., hard wares;
d. Or some may combine production of latrines, transportation, installation, operations and maintenance services

3.3   Summarize this session by reminding participants that:

♦ Every business starts small by responding to an opportunity — a need
for product or service.
♦ To overcome start-up challenge, entrepreneur's motivation is key
♦ But grow and expand, an entrepreneur requires capital, good
marketing skills and partnerships with others.

Step 4: 5 minutes

Mapping Partners in Your Value Chain

4.1   Start this section by reminding participants of this wise African saying that is relevant
to enterprise: "If you want to go fast, go alone. But if you want to go far, go with
others'.

4.2   So in business, partnerships are important. Every entrepreneur must therefore know
who the right partners for his/her business are. An important tool to use in mapping
strategic partners is called a value chain map — relationships between entrepreneurs
and supporting partners work together to meet the needs of a consumer.

♦ **Example:** For Kulavi to construct a SAFI latrine for household, he depends on NGOs
and Public Health officers to sensitize households about the toilet and train him
on construction. He also depends on a hardware shop for materials like cement,
nails or iron sheets. From the hardware, he will depend on a local transporter to
take materials to site before constructing a full toilet for household. This chain
of actors comprise a value chain,

♦ We can draw this relationship using a chart as follows:

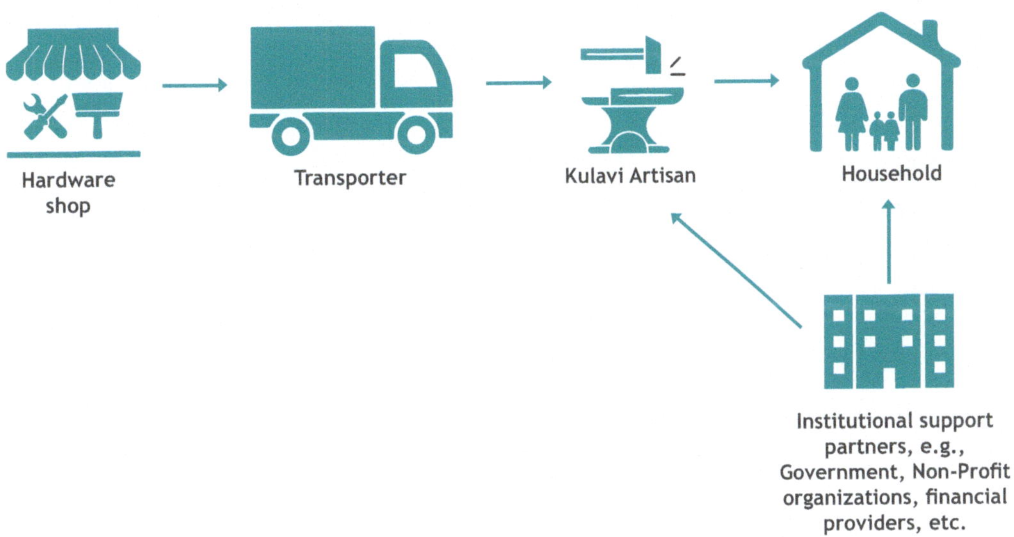

Exercise:

Ask participants to brainstorm and identify

    a. Which local actors provide products or services related to what your business wants to offer?

    b. Which suppliers are needed to start or improve my business?

    c. Where are they located and how does it affect my business?

4.3   Summarize this session by reinforcing the following points.

- ♦ Know your business competitors and those who sell similar products.
- ♦ Map out suppliers and establish relationships with players that add value to our business.
- ♦ Other partners, e.g., Government institutions may formulate laws, rules and practices that apply in starting and running a business while Not-for Profit organizations may provide business development services, capacity development trainings or financing for small businesses.

Exercise

Can you map the activities involved to deliver value of a SATO product to a customer in your remote village, and the added cost at every stage?

Hint: Ask yourself who manufactures the SATO and at what cost? Who picks from the factory at what cost? Who transports at what costs, who retails/wholesales and at what cost? Are there any other players in this business and what are their roles?

4.4 Summarize this session by highlighting the different products or services offered and point out the market served on the sanitation value chain. Lastly, ask participants reflect on the two questions:

Does your business add any value? What is your role?

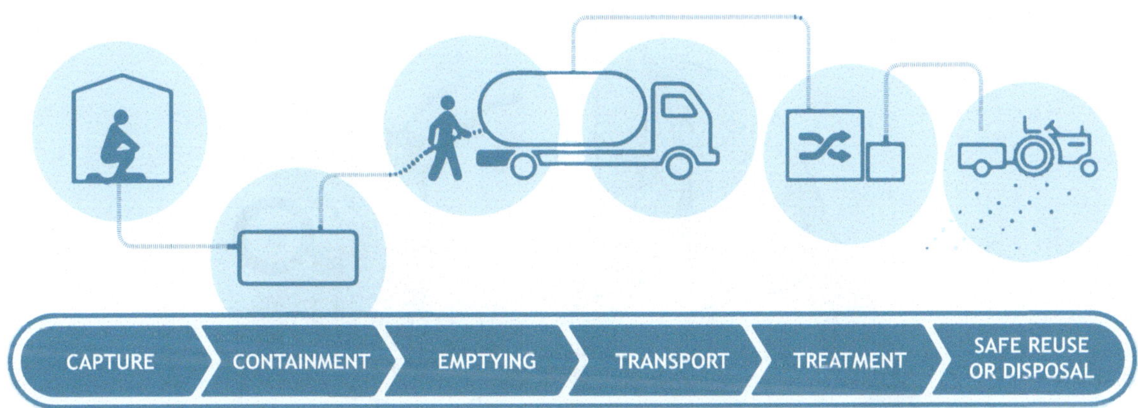

Example: Conceptual model of a conventional sanitation value chain.

Step 5: 5 minutes

Learning Assessment
Training Materials: Handout 1.4

Using Handout 1.4, ask participants to assess each of the three sessions above by ticking on the appropriate box as relevant. Illustration in a tick box:

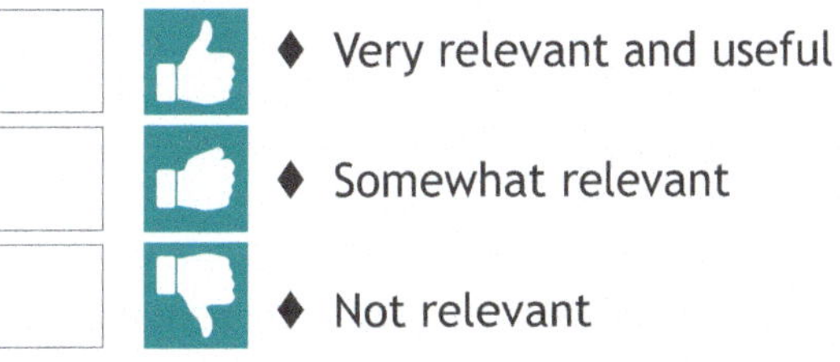

Handout 1.1

Concept Definitions

What is an enterprise?

An enterprise is another word for "business."

Who is an entrepreneur?

An entrepreneur is someone who engages in a business activity/enterprise. This is someone who takes initiative to organize, set up a business to benefit from an opportunity, and decides how much of a good or service to produce or sell.

What is (social) entrepreneurship?

The capacity and willingness to organize, start and manage a business venture that solves a community problem and earns a profit doing so.

What is sanitation entrepreneurship?

The concept that encompasses transforming challenges, ideas and visions to "new business ventures or expansion of an existing business" in sanitation.

The Case of a Successful Sanitation Entrepreneur

Timothy | Sanitation Retail Entrepreneur

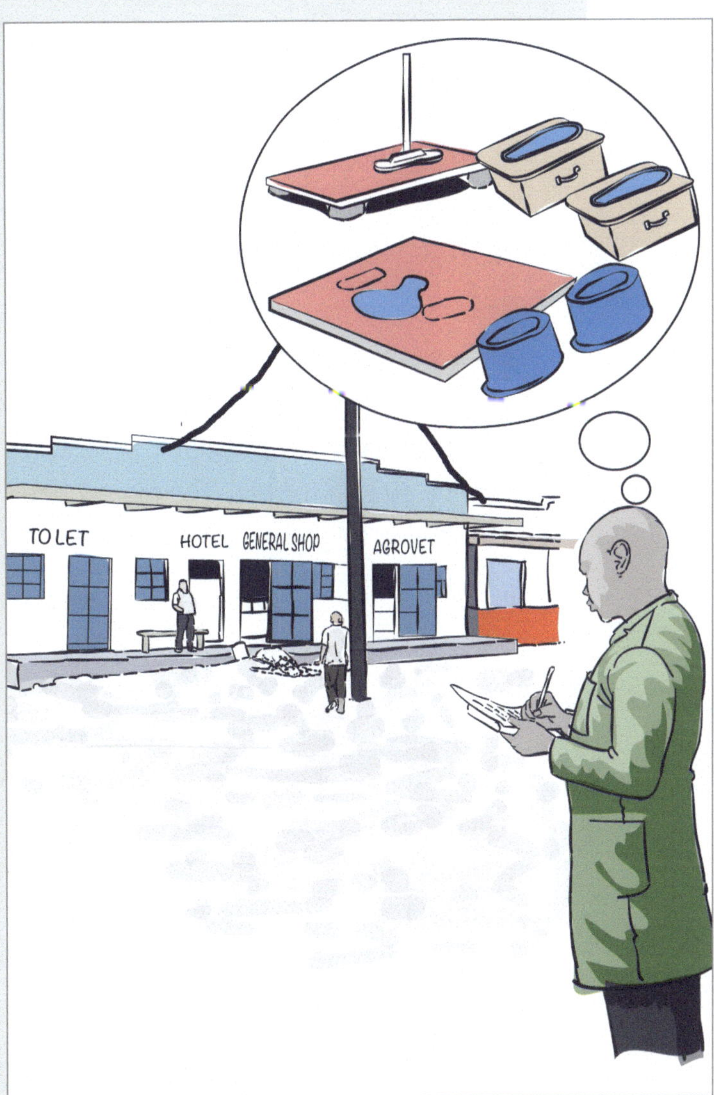

57-year-old Timothy from Kakamega county works as a community health volunteer in charge of his neighborhood unit. Apart from working with the Ministry of Health, he also practices farming and is passionate about carpentry and masonry.

The USAID-Kenya Integrated Water, Sanitation and Hygiene (KIWASH) Project launched sanitation and hygiene-promotion activities in collaboration with the

Ministry of Health in Kakamega County. Timothy played a lead role in educating his fellow community members on the importance of access to sanitation and good hygiene practices. As he went about his roles, Timothy realized the lack of improved sanitation options and technical skills in construction of improved latrines.

In 2018, Timothy was lucky to be selected as a participant in an artisans training organized by KIWASH. "KIWASH trained me on how to construct SAFI latrines and install SATO products. This was an eye-opening opportunity for me," says Timothy. "Immediately after the training, I put up a SAFI latrine demo site and purchased SATO products from CIL Africa using my savings, this was my marketing strategy." In addition, Timothy constructed a complete SAFI latrine at the entrance of his home. "Thanks to this demo site, I receive many visitors and clients who are interested in constructing improved latrines."

Timothy also markets the products in the local market and goes door-to-door to reach more people. "My marketing slogan revolves around elimination of flies, smell and ease in cleaning which has continuously won the hearts of many." The business has not only improved hygienic practices in my community, but aso has created a new revenue stream for me. The money I get from the latrine construction venture has enabled me to pay school fees for my children at the university," he proudly notes.

In order to access more income to expand his business, Timothy partnered with a local women's group after sharing the business idea. The group pools resources together through the sale of shares and loans among members. The group has so far invested USD120 in purchasing SATO products which KIWASH doubled through a matching grant. "Despite the venture slowly gaining popularity, my records indicate that I have so far installed 59 SATO products and constructed two SAFI latrines," notes Timothy. The gradual uptake of improved sanitation products can be attributed to low income in the area. There are a number of clients wtho have defaulted payments despite having installed SATO products for them.

Notwithstanding, Timothy has remained committed in promoting and marketing improved sanitation technologies. He uses his prototype SAFI latrine models and SATO models for demonstration and exhibition in county events to market the products. In addition, he also makes cleaning brushes for SATO products that retail at USD 1 and uses social media to promote his business.

Important Elements of a Sanitation Enterprise

Market Opportunity

Every business requires a market: buyers willing and able to buy the product or service. As a start-up sanitation entrepreneur, you need to understand what your target customers want and how much they will pay for it. Timothy knew that most villagers, landlords, and hoteliers liked the SATO latrine. This was the opportunity. The marketing session will explore these concepts more.

Vision

Every successful entrepreneur in the world today had a great vision for themselves and their business. They pursued market opportunities, sought ways to grow, and transformed their lives and those of their families and communities.

Plan

A marketing opportunity and a vision without a plan of action remain just wishes! To be successful in any business, every entrepreneur must have a roadmap or strategy to succeed. The types of plans could include:

♦ Business plan or strategic plan

Capital

Entrepreneurs need capital at every stage of their business growth: capital to start the enterprise, working capital to meet any cash flow shortages during operations, and capital to expand operations. Every enterprise must understand capital requirements and sources.

Cashflows

Cash is the lifeline of any business! Entrepreneurs receive cash from sales and

in form of loans, and pay out money for operating costs, purchases, and repaying loans. Any shortage in cash means that the enterprise is in distress. We shall learn more about cash flow management and accounting during the financial literacy session.

Motivation

The desire to start, grow and lead a successful enterprise. It is the driving force that every entrepreneur needs to overcome obstacles and succeed.

Profit

A financial gain, especially the difference between the amount earned and the amount spent in buying, operating, or producing in an enterprise.

Handout 1.4

Participant Evaluation Form

Ask participants to provide feedback on each section by signaling with thumbs up, thumbs flat or thumbs down. Mark with ticks or numbers participants' feedback for each session in the boxes below.

Session	Very relevant and useful	Somewhat relevant	Not relevant or useful
Entrepreneurship Concepts			
Characteristics and qualities of an entrepreneur			
Setting up a sanitation enterprise			

Handout 1.5

Timetable

Course:	Business Skills Training for Sanitation Entrepreneurs		
Module:	Entrepreneurship		
Session: An Entrepreneurs Mindset		Venue:	Date:

Time	Activity	Facilitator
5 minutes	Ice breaker and session introduction	
5 minutes	Key Concepts: *Enterprise* and *Entrepreneurship*	
15 minutes	Characteristics of a sanitation entrepreneur *Case Study*: *Timothy Kulavi* Key attributes of entrepreneurs	
5 minutes	**Break**	
15 minutes	Things to consider when starting a sanitation enterprise *Experience Sharing*	
5 minutes	Session Assessment	
	End	

Module 2

BASIC MARKETING SKILLS

Session Plan

Session Title	Basic Marketing for Sanitation Products
Time	55 minutes
Objectives	At the end of this session, participants should be able to:
	◊ Identify their customers and how to expand their customer base
	◊ Articulate their own selling and marketing strategy
Contents	◊ Understanding their customers
	◊ Selling tactics for different customer segments
	◊ Developing your personal marketing plan
Trainer Resources	◊ Illustrations 2.1, 2.2, 2.3, 2.4
	◊ Handout 2.0
	◊ Participants workbook
Methods	◊ Picture story
	◊ Discussion
	◊ Question and answer discussion
Venue	In-person in a public space (church, school, etc.)

Session Guide

Defining Key Terms in Marketing

1.1   Welcome participants to the marketing class, which builds on Session 1, Entrepreneurship.

Define the terms:

♦ "Customer" as the person or group who want the product or service and are willing to pay it.

♦ "Marketing" as: Activities undertaken by a company to promote the buying or selling of a product or service.

♦ Social marketing and commercial marketing as used in sanitation enterprise. Use handout 2.1

1.2.   Explain to participants why social marketing skills are necessary for selling toilet products and services in communities.

Understanding Your Customer
Training materials: Illustrations A and B

A   B

2.1 Look at Illustrations A and B of two SATO entrepreneurs on next page.

Ask participants the following questions:

- ◆ What do you see in illustration A? What do you see in illustration B?
- ◆ Between entrepreneur in A and B, who is likely to sell more? Why?

2.2 Ask participants to reflect on the following questions.

- ◆ Which people or groups in the community are interested in my product or service? The list may include: farmers, churches, local hotels, landlords, teachers, etc.
- ◆ Why do they like or dislike my product? (color, size, functionality, status, etc.)
- ◆ Are they willing to pay for it? When do they want the product — regularly or seasonally?
- ◆ Where should my products or services be available? (Location of sales points)

2.3 Summarize the session by reinforcing the following points:

- ◆ The customer is king in business. They only pay for what they want. Highlight that customer preferences vary by gender, age or income status, etc.
- ◆ **Marketing** is communicating information about your product or service with features, benefits and solutions tailored to clients. This differs from selling, which is the actual act of persuading or exchanging the goods for money.
- ◆ **Customers in most cases are impulse buyers.** People make decisions to buy a product or service if they can connect to your sales message (mainly attributes and functionality of the product or service). In these illustrations, entrepreneur A's message connects with potential customers.

C   D

2.4 Looking at Illustrations **C** and **D** above, ask participants why different customers might buy SATO products? Potential answers may include:

- **Price:** affordable to the customer
- **Functionality:** eliminates flies, bad smell and easy to clean. SATO stool solves problem of elderly people having difficulty squatting
- **Aesthetics:** attractive to the eyes
- **Availability:** the product is available on demand
- **Durability:** product may last for a longer time

Note to the trainer:

If participants are trading with other latrines, ask similar questions in 1.6

2.5 Discuss with participants on their experience with good/bad customer service.

Reflection task:

Imagine where you usually shop in your local center or local market. How does your shopkeeper or seller treat you at their place of business? Would you continue going there if they keep you waiting in the queue for long? How about if they were to be rude to you? What if you had a complaint about their product or service, and which took long abit to be addressed, but they apologized and rectified.

♦ Would you go back there again?

2.6 Wrap up the session by letting participants know that:

♦ Each toilet product has its own market segment

♦ Customers buy product/services based on attributes. When these attributes are bundled together, it becomes your sales pitch

♦ A good sales pitch (marketing promotion) compels a potential customer to buy your product or service

♦ Always keep in mind what attribute to emphasize more for each potential client

♦ Entrepreneur must have a unique selling proposition for each target market

♦ Every customer needs to feel appreciated and valued. They need to feel that you as the seller of a good or service is honest, and you give them the best product or service.

Step 3: 45 minutes

Market Positioning and Selling Propositions

3.1 With the aid of a flipchart, ask participants to list to discuss and list the main attributes of each toilet product — SATO, SAFI and VIP latrines; and which customer would buy the product. Attributes may be in categories like: Functionality, Components, Cost, Beauty, Durability, etc.

3.2 Define the term "positioning" to participants. Then guide each entrepreneur to follow three steps to position their product or service:

♦ State potential customers and the competition

♦ Segment the market and select their target customers

♦ Identify the position of the product or service

Positioning refers to the image you would like the customers to have about your product. Without a clear position, you would not know what to do with the Marketing Mix (the 5Ps that we shall learn in the next session). This is similar to identifying a destination before choosing the way to get there.

A good position should give customers a strong reason to buy your product. Therefore, it should:

- ♦ Respond to a need that customer's value highly
- ♦ Differentiate your product or service from your competitors
- ♦ Make sure that you can deliver what you promise

3.3 Inform participants that selling a toilet product may require special skills than selling a consumable product like a loaf of bread for various reasons:

- ♦ An improved toilet can be costly and may not be a priority to households.
- ♦ Benefits of an improved toilet may not be real to many low-income households
- ♦ The art of selling must be well thought out and convincing.

3.4 Case study – Knowing What to Sell to Your Customer.

LUCILLA, Entrepreneur

Lucilla, a 40-year-old lady, has a pit latrine with a mud slab, wall and grass thatched roof. She lives with her family of six, including her aging mother-in-law in a rural village in Rarieda. She is employed as an Office Assistant at the County Government of Siaya and earns a salary of Kenya shillings (KES) 60,000.

Lucilla wants to have a hygienic latrine. The safety of her young children is important to her. She wants a modern latrine like the one she saw in Kisumu recently; one that will make her be respected at her village. She believes that a good latrine enhances one's own dignity and respect. But at the same time, she believes such latrines are very expensive and unaffordable due to her limited salary.

Exercise:

Role Play: Ask a woman volunteer from participants to act as Lucilla. Divide participants into two groups. Ask each group to identify a type of toilet, which they are familiar with, its attributes and convince Lucilla for 5 minutes on why she should get their toilet. Let the two groups choose different toilet technologies.

3.5. Ask participants to brainstorm and share their thoughts on why Lucilla may like a latrine but not own it? Do all people own latrines? If not, why?

Conclude by saying:

Many people want to own latrines but are unable to pay for latrines on cash upfront terms. By offering financing options, entrepreneurs can increase the number of latrines they sell.

3.6. Ask participants to brainstorm and identify financing options that they can offer to customers.

Tips: Participants may highlight;

- ◆ Payments in monthly installments with or without interest
- ◆ Educating customers on how to borrow loans through local village savings and credit groups

3.7. Summarize the session with the following points;

♦ Each toilet product has its own market segment.

♦ Product attributes, when bundled together, determine your market position.

♦ A good proposition gives customers a strong reason to buy your product or service.

♦ Always keep in mind how they want clients to think of their product/service.

♦ Every entrepreneur must have a unique selling proposition for target market.

♦ Supporting consumers financing requires an entrepreneur to keep very good records and create policies around repayment.

♦ Entrepreneurs need to invest in a system that vets each customer's credit worthiness and facilitates efficient follow-ups with loanees to pay off their debts

Step 4: 15 minutes

Developing Your Personal Marketing Plan
Training materials: Participant workbook

4.1   At this point, participants may have a mental picture of how to market and sell their products or service. Individually or as a group, ask them to answer this question:

♦ What would you do differently to increase your sales after today's session?

♦ Allow participants to write down in their workbook a few tactics and share with the group.

♦ Emphasize the need of having a clear sales pitch to each customer segment.

Step 5: 5 minutes

Learning Assessment
Training materials: Handout 2.2

Using Handout 2.0, ask participants to assess each of the three sessions above by ticking on the appropriate box as relevant.

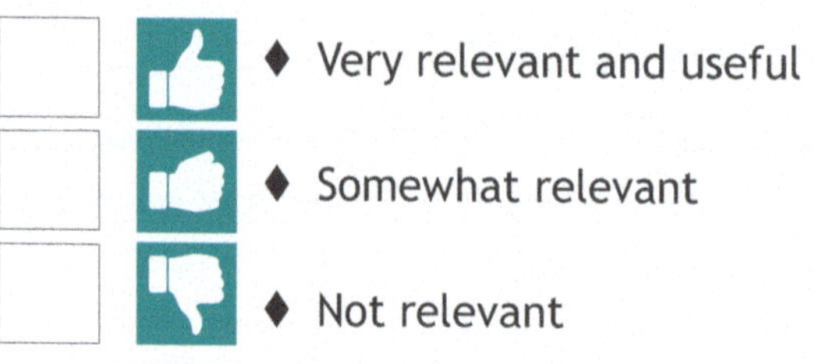

Handout 2.1

Participant Evaluation Form

Ask participants to provide feedback on each section by signaling with thumbs up, thumbs flat or thumbs down. Mark with ticks or numbers participants' feedback for each session in the boxes below.

Session	Very relevant and useful	Somewhat relevant	Not relevant or useful
Understanding the customer			
Customer segments			
Developing your personal marketing plan			

Handout 2.2

Timetable

Course: **Business Skills Training for Sanitation Entrepreneurs**
Module: **Basic Marketing**

Session:	Venue:	Date:

Time	Activity	Facilitator
5 minutes	Ice breaker and session introduction	
15 minutes	Understanding your customer	
20 minutes	Customer segments	
15 minutes	Developing your personal marketing plan	
5 minutes	Learning assessment	
	End	

Module 3

BASIC FINANCIAL LITERACY

Business Skills Training for
**Rural Sanitation
Entrepreneurs**

Session Plan

Session Title	Financial Literacy for Sanitation Entrepreneurs
Time	2 hours, 30 minutes
Objectives	At the end of this session, participants should be able to:

At the end of this session, participants should be able to:

◊ Understand the concept of financial literacy and behaviors of financially literate people

◊ Appreciate the importance of saving and borrowing for enterprise

◊ Understand cost of product/service and pricing for profit

Contents

◊ Basic Financial Literacy Terms

◊ Saving

◊ Borrowing

◊ Costing & Pricing

◊ Money Planning

Resources

◊ Flip chart

◊ Marker pen

Methods

◊ Case studies

◊ Metaphors

◊ Question and answer

Note to trainer

The depth and scope of this module depends on the numeracy levels of the participants, size of their enterprise(s) and competency of the trainer on the subject matter

Basic Financial Literacy Terms

1.1 The session introduces sanitation entrepreneurs to the concept of financial literacy. It will develop their ability to make informed and effective decisions regarding how to save and borrow money and use the money to grow their business.

1.2 Read the story below to trainees and discuss the questions that follow:

A Tale of Two Hardworking SATO Entrepreneurs

Two start-up sanitation entrepreneurs, Edna and Liz, are community health workers from Biticha CHU in Nyamira county. In January 2019, they received a stipend of KES 20,000 each from the county government for their work in 2018. Edna decided to save KES 5,000 in their Village Savings and Loans Association (VSLA) group, which she intends to use to start a SATO retail business. There is growing demand for SATO products after the KIWASH project's work in her village. She used another 10,000 to start a poultry business. She bought 50 one-month-old chicks and will sell them as mature chickens in January 2020 to pay her children's school fees. She gave her husband KES 2000 to travel for an interview in Nairobi and saved the remaining KES 3,000 in case of emergencies.

Liz decided to invite her sisters to celebrate her milestone as a health volunteer contracted by the county. She used KES 8,000 to buy foodstuffs and drinks for the celebration and new clothing for herself and her children. She then saved KES 1,000 in their VSLA group to start SATO business because it was mandatory for each member. Further, she sent KES 5,000 to her mother to buy some new clothes for an upcoming wedding. She then gave the remaining KES 6,000 to her husband for his own use.

Discussion questions:

1.   What characteristics do Edna and Liz possess?

2.   Who among the two do you admire and why?

3.   Who among the two has good money habits and why?

4.   Who will have a better future and why?

1.3   Experience sharing on money management by participants:

♦ Can you identify one time when you did a good job handling money? How did it help you?

♦ Can you identify a time when you used money carelessly, perhaps because you had limited knowledge of financial literacy?

1.4   Summarize the session with the following points:

♦ **Financial literacy** can be defined as a set of skills and knowledge aimed at helping one make sound financial decisions. This includes knowledge about how to make money, how to plan for expenses, saving and borrowing, and even investing.

♦ Financial literacy is all about money management techniques. Financially illiterate people LOSE money through unnecessary personal and business expenditures.

♦ Financially literate people make WISE money decisions. They GROW their money by saving or investing in viable enterprises.

Saving

This session introduces participants to the practice of saving, its importance, and different methods and ways of saving for personal and business growth.

2.1   Explore the concept of saving using the following story:

The Ant and the Grasshopper

One beautiful day, Grasshopper was hopping about in a green field of grass, singing and whistling and making music from his heart. It was a good day for Grasshopper, after all, as there was plenty of grass for him to eat.

In the same green field Ant was busy gathering grass and storing as much as he could in his ant hill. Grasshopper laughed at Ant for wasting time gathering and storing grass instead of enjoying and making merry. Ant cautioned Grasshopper that he should store up food because when the cold season comes, water and snow will cover the earth and there won't be any green grass in the fields to eat. But Grasshopper ignored the warning and continued with merry making.

When the cold weather came, the grass dried away and food became scarce. Ant had stored up enough food to make it through the season with plenty to eat. But because Grasshopper did not store up any food during the warm season, he was out in the cold, hopping from one place to another in search of grass to eat. He could not find any and died.

    Discussion Questions:

- What was Ant saving for? [Food for the cold/rainy season]
- What did Ant give up? [Time for playing or relaxing]
- What was Grasshopper's opportunity cost? [No food for winter]
- What do we learn from the story? (The need to save when there is plenty, saving is not enjoyed by everybody, saving often means sacrificing)
- How was Ant's behavior beneficial?

2.3   Reflection on personal experiences:

- What have you had to give up in order to get something else?
- What do you need to do to develop the discipline of saving?
- What could you invest your savings for?
- How do you currently save? Do they save for sanitation enterprises?

2.4   Summarize the session by reinforcing the following points:

- What Ant was doing is similar to saving.
- When we earn money, the first expenditure should be savings. Pay yourself FIRST. Do not SAVE what is left after SPENDING!
- You can save in your VSLA, Savings and Credit Cooperative (SACCO), bank or buy assets that appreciate in value like livestock (chicken, goats, cow) or land.

NOTES

What is saving?

Saving is putting money aside for future use. When you keep your money somewhere for a purpose, then you are practicing saving. Saving can also be defined as a portion of income not spent on current expenditures. Saving is not done once but over a period of time. You may have to sacrifice current luxuries to save for a better future.

What is a savings plan?

A savings plan is a tool for handling money to meet short, medium, or long-term financial goals.

How to Make a Savings Plan:

- Define a savings goal — an amount you want to save.
- Set a target date to reach your goal.
- Divide the goal (amount) by the number of months until your target date to determine your monthly savings requirement.

- Determine your monthly earnings, and then subtract your fixed costs to determine how much extra money you have to save each month.

- Do you have enough extra money after fixed costs to reach your goal? If not, identify which expenses you can cut back on, and reallocate this amount to your savings.

- Decide where you will save your money. These places may include a Savings Group, a SACCO, and Bank or in an asset.

Step 3: 30 minutes

Borrowing

This session introduces participants to the concept of borrowing, why a sanitation enterprise should borrow and the key issues to consider before and after borrowing.

3.1 Read through the scenario below and let participants answer the questions that follow.

The Kanyimach Self Help Group SATO Loan Experience

The 12-member Kanyimach Self Help Group in Rongo, Migori County started in 2016. The group has accumulated KES 50,000 in savings that revolves through its VSLA scheme. In March 2019, a KIWASH sanitation specialist trained them on the concept of a Sanitation Revolving Fund (SRF), where members save to buy SATOs to sell. The group developed a credit facility (loan in-kind) for its members, where they take SATOs on credit and repay in cash on monthly basis. However, at the time of share out in September, 5 members failed to repay their SATO loans. They were not attending group meetings, and their savings could not cover for the credit. The group resorted to forceful recovery. This brought bitter conflict though this was later on resolved. After the share out, the group had no capital to sustain the SATO business.

The group chairperson proposed that they go for a loan of KES 50,000 from their bank to buy more SATOs and meet growing demand.

3.2 Discussion Questions:

(a) What measures should this group put in place to avoid member loan defaults?

(b) What advice would you give the Kanyimach Self Help Group on the proposed bank loans? Is it a good idea or a bad idea?

(c) What do you think the bank will consider before giving the group a loan?

3.3 Summarize the session by reinforcing the following points:

b) There are pros and cons of using your own money vs. borrowed money

- You are freer when you use your own money.

- A loan costs money (interest).

- Some people lose property when they fail to repay the loan.

c) However, by borrowing, you get a greater lump sum faster than you might when using your own capital (more money quickly than if you rely on your ability to save little-by-little).

d) Loans are costly and one should be careful when borrowing from commercial sources. The cost may be direct or indirect as explained below:

- **Direct cost:** Money you pay to the bank for the loan, including interest, fees, insurance, legal fees and late penalties. Some of these costs (fees, insurance) are paid upfront before you get the loan.

- **Indirect cost:** Expenses you may have to pay because you have the loan, such as bus fare to go to the bank to apply and negotiate the loan.

e) Governments and non-governmental organizations may also give loans or grants to small entrepreneurs willing to start or scale-up their businesses.

Costing and Pricing Products and Services

Handout 3.1 - Case of Nthongoni SATO product enterprise

This session helps participants calculate the full cost per unit of sanitation products or services sold to a client.

4.1   Start the session by discussing the concepts of VARIABLE COST and FIXED COST.

- ♦ **Variable costs:** Costs that change with number of products sold.
- ♦ **Fixed costs:** Costs which remain the same irrespective of products sold.

4.2   Study the scenario's on next page and calculate per unit cost of the sanitation products.

Costing and pricing a SATO product retail enterprise

After their KIWASH training, all 37 members of Nthongoni Community Health Unit (CHU) in Kibwezi, Makueni County agreed to start a SATO retail enterprise in November 2018. They required capital of KES 23,135. This paid for supplies, transport to and from Nairobi, and airtime for the chairman. The group agreed that each member would make a contribution of KES 355 (total KES 13,135) in cash. To meet the required budget, a further KES 10,000 was obtained from group savings in the bank.

Less than one year later, the group now sells on average 60 SATO pans and 10 SATO stools each month. SATO pans come in boxes of 10 units, and each box costs KES 4,500. SATO stools come in boxes of

five units, and each box cost KES 4,000. The group's monthly revenues are now KES 38,000. The group sells the products at KES 6,000 and KES 5,000 per box, respectively. They make direct purchases from the manufacturer in Nairobi and keep proper records of their stocks and costs. The group's cashbook shows a record of the following costs for orders made and sold:

- MPESA charges of KES 220 when purchasing the SATOs
- Transport per box of SATO pan is KES 150 and KES 200 for a box of SATO stools
- The chairman receives KES 500 worth of airtime to coordinate purchases and delivery of products to the group every month
- The group has an arrangement with a local shop to store its stocks. The shopkeeper charges a flat fee of KES 400 for storage.
- A commission of KES 20 and 50, respectively, is paid to members for each unit sold.
- It pays a local artisan KES 300 to install a SATO pan and KES 200 for a SATO stool. But clients pay the group KES 400 and KES 300 for installation, respectively.
- The group meets monthly to reconcile records. The budget for teas and snacks for the meeting is KES 300.
- The group offers its customers the product and installation service. For every unit, the group expects to retain a profit of KES 150 for the SATO pan and KES 200 for the SATO stool.

Exercise:

- Which of the above costs are VARIABLE and which are FIXED?
- Calculate per-unit VARIABLE cost and per-unit FIXED cost of each product.
- Calculate the price per unit at which Nthongoni CHU
- Compare the per unit cost and the per unit sales price. Is the group break-even (making profit) with this price?
- What is your advice to the group?

Solution:

- Guide participants through these three steps, which will help answer each of the questions.

Step 1

Separate variable cost from fixed costs as follows:

- ◆ **Fixed costs:** Chairman's airtime, storage and meeting costs.
- ◆ **Variable costs:** cost for purchases, MPESA money transfer, transport, installation and commissions.

Step 2

- ◆ Remember that we have two types of products: SATOpan and SATOstool.
- ◆ Calculate total variable cost for each product. Then calculate per unit variable cost by dividing with the number of each unit.

Step 3

- ◆ Sum the per unit variable and fixed costs per product to get total unit cost.

The result will be as shown in Table 1 here.

Cost Item	SATOpan	SATOstool
No. of boxes	6	2
No. of units	60	10
Purchase price per bo	4,500	4,000
Variable Costs		
Purchases	27,000	8,000
MPESA charges	170	50
Transport	900	400
Installation	18,000	2,000
Commission	1,200	500
Sub-total (VC)	47,270	10,950
VC per unit	788	1,095
Fixed Costs		
Chairman's airtime	429	71
Storage	343	57
Meetings	257	43
Sub-total (FC)	1,029	171
FC per unit	17	17
Total Unit Cost	805	1,112

Table 1. Nthongoni CU: Per Unit Variable and Fixed Cost

Step 4

- ◆ Compare per unit cost from step 3 and the per unit selling price. Then provide your advice to the Nthongoni CU.

Total Unit Cost	805	1,112
Per Unit Sales Price	600	800

Note to the trainer:

The following case and exercise is relevant to participants in construction and trading of SAFI or ventilated improved latrines.

Handout 3.2 - Case of Vaele Women Group

Costing and Pricing of a SAFI Latrine

Vaele Women Group in Kakamega county has twenty members. It comprises of CHV's and a few artisans who were trained on SAFI latrine production and installation. The group heard about SAFI retail business opportunity from a USAID funded KIWASH project that had donated a SAFI latrine production mold for use by the local community. In January 2019, members came together and agreed to expand their business and start SAFI latrine production and retail. The group calculated the cost of producing one SAFI latrine, considering the costs of materials used and other costs incurred related to labor, time spent in manufacture of one latrine and transportation cost of materials to

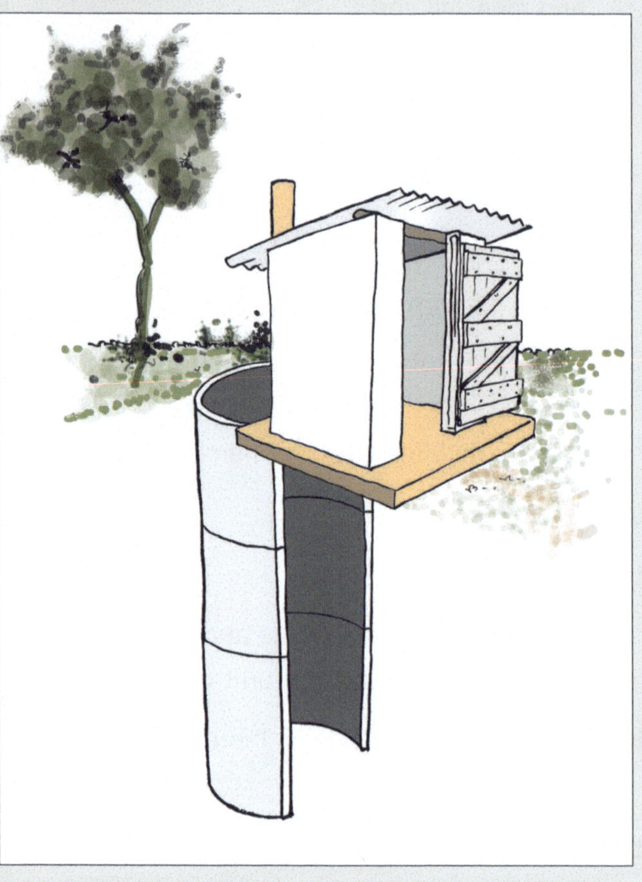

the manufacturing site. Table 1 shows the breakdown of costs considered. The total cost of producing one SAFI latrine to Vaele group was KES 14,850 and the group projected to sell at least five SAFI latrines a month. Based on these projections, group members agreed to invest from their group savings an initial start-up capital of KES 75,000. They agreed to set a profit margin of 10% for each latrine produced. Since July 2019, Vaele group has sold six SAFI latrines and increased their sales revenue to KES 98,010. From its cash book records, the group;

♦ Pays a commission of KES 500 to every member responsible for a unit sale

♦ Charges the household an extra KES 3,000 to provide a local artisan who can construct a SAFI latrine but Vaele group pays the artisan 2,500 per unit of construction

♦ Pays an annual business permit fee of KES 1,000

The group meets monthly to reconcile records. The budget for tea and snacks for the meeting is KES 300

4.3. Using the template provided (Hand out 3.1) at the end of this module, calculate the production cost for an improved latrine in your local area. Ask participants to cost the production, overheads (electricity, etc.) as well as factors such as labour and depreciation of equipment.

Ask them to set a profit margin and selling price. Remind participants that the profit earned by business may not be retained in its entirety. The business may need to pay taxes to the government or local government

4.4. From the exercise above, which cost can be classified as FIXED or VARIABLE cost? List in the table provided in your workbooks.

4.5. Assuming that some clients live far away from Vaeles production site and require latrines with varying number of rings (1–4). Would you predict a price change? Draw the attention of participants and discuss the importance of the worksheet below.

Distance relative to Vaele Groups production site in kilometres (km)	Number of rings required/latrine pit			
	1 Ring	2 Rings	3 Rings	4 Rings
Transportation to villages < 10 km				
Number of km over 10				
Added fee per km				
Transportation to villages > 10 km				
Total Transport Cost				

4.6. Summarize the session with the following points:

♦ Every entrepreneur must understand the full cost of their business.

♦ For pricing, calculate per-unit cost of delivering your product or service.

♦ Add your expected mark-up (or profit margin) to get the price to charge your customer.

♦ Classify variable and fixed costs to enable you control your costs.

♦ Increase profit margin by reducing on per unit cost of production. Reduce on per unit cost of production by increasing quantities produced.

Money Planning

This session introduces participants to good money planning habits and tools for a successful enterprise.

5.1. Ask participants what happens when a tree seedling lacks water. It dries up and dies. And what happens when a new SATO retail business runs out of money? It can fail. When any business runs out of money, it's a big problem! You cannot buy any new stocks for your business, you cannot meet your costs like transport, airtime, wages to people who work for you, or pay any debts falling due.

But when a tree seedling has enough water, it can grow and become a huge tree in the future. In the same way, keeping enough cash in your business will allow it to run well and grow.

5.2. Every entrepreneur must plan ahead to avoid running out of money. A Money Plan (or Cash flow Budget) is every entrepreneur's important money management tool.

A Money Plan helps you to:

 ♦ Figure out how much money you need to start or to keep your business running until it makes enough money to sustain itself; and

 ♦ See how much cash (or shortfall) you will have at end of every business period (week, month or quarter, etc).

 ♦ Good money planning increases the enterprises likelihood of accessing commercial loans or credit

5.3.   Reflect on the above case of Nthongoni CHU and answer the questions below.

Questions:

Assume that the group does not want to run short of products at any given time to avoid customer complaints. It therefore utilizes any profits to increase its purchases by one box of SATO pans every month for the next six months. Sales and fixed costs per month remain unchanged as discussed in the case.

- ◆ Will Nthongoni CU have enough cash to meet operations?
- ◆ How much cash surplus (or shortage) will the group have?
- ◆ What is your advice to Nthongoni CU on how it can improve its cash position?

Hint: To answer this, we require to prepare a Money Plan, a forecast of MONEY IN from sales and MONEY OUT for purchases and expenses. Guide participants to use the templates provided in their work-books to prepare their Money Plans following the step-by-step process below.

How to Develop a Money Plan

First indicate on the quantity of SATOs to be purchased and Quantity to be sold for every product during the six-month period.

Table 1: Illustration of Nthongoni Stock

ITEM		MONTH					
		1	2	3	4	5	6
No. of boxes to purchase	SATO pan	6	7	8	9	10	11
	SATO stool	2	2	2	2	2	2
	Total	**8**	**9**	**10**	**11**	**12**	**13**
No. of boxes to sell	SATO pan	6	6	6	6	6	6
	SATO stool	2	2	2	2	2	2
	Total	**8**	**8**	**8**	**8**	**8**	**8**
Stock balance (boxes)		0	1	2	3	4	5

◆ On each row of the MONEY IN, the group will receive cash for all eight boxes. It will also receive money from installation fees from its customers.

◆ On the MONEY OUT section, the group will also pay out money to purchase, transport, store and install the products.

Table 2: Illustration of Nthongoni Money Plan

		MONTH					
MONEY IN		1	2	3	4	5	6
	Sales	46,000	46,000	46,000	46,000	46,000	46,000
	Installation fees	30,000	30,000	30,000	30,000	30,000	30,000
	Total	**76,000**	**76,000**	**76,000**	**76,000**	**76,000**	**76,000**
MONEY OUT	Purchases	35,000	39,500	44,000	48,500	53,000	57,500
	MPESA charges	220	220	250	250	270	270
	Transport	1,300	1,450	1,600	1,750	1,900	2,050
	Chairman's Airtime	500	500	500	500	500	500
	Storage	400	400	400	400	400	400
	Installation fees	20,000	20,000	20,000	20,000	20,000	20,000
	Meetings	300	300	300	300	300	300
	Commissions	1,700	1,700	1,700	1,700	1,700	1,700
	Total	**59,420**	**64,070**	**68,750**	**73,400**	**78,070**	**82,720**
BALANCE	**START= 38,000**	**16,580**	**11,930**	**7,250**	**2,600**	**(2,070)**	**(6,720)**

Note for the trainer: Use exercise on 3.4 if training SAFI latrine enterprises

5.4.   Reflect on the above case of Vaele CHU and answer the questions below.

Questions:

Assume that the group does not want to run short of products at any given time to avoid customer complaints. It therefore utilizes any profits to increase its purchases by producing one SAFI latrine every month for the next three months. Sales and fixed costs per month remain unchanged as discussed in the case.

◆ Will Vaele group have enough cash to meet operations?

◆ How much cash surplus (or shortage) will the group have?

◆ What is your advice to Vaele group on how it can improve its cash position?

Hint: To answer this, we require to prepare a Money Plan, a forecast of MONEY IN from sales and MONEY OUT for purchases and expenses. Guide participants to use the templates provided in their work-books to prepare their Money Plans following the step-by-step process below.

How to Develop a Money Plan

Table 2: Illustration of Vaele group stock plan

Vaele Groups 3 Months SAFI latrine Stocks							
	Month	1	2	3	4	5	6
No. of SAFI latrines to produce	Total	6	7	8			
No. of SAFI latrines to sell	Total	6	6	6			
Stock balance (unsold SAFI latrines)		0	1	2			

- On each row of the MONEY IN, the group will receive cash for all six SAFI latrines (assumed selling price was KES 20,000/unit) in cash. It will also receive money from latrine construction fees from its customers.

- On the MONEY OUT section, the group will also pay out money to artisans, business permit fee and meeting costs

Table 3: Illustration of Vaele group Money Plan

		MONTH					
MONEY IN		1	2	3	4	5	6
	Sales	120,000	120,000	120,000			
	Artisan Construction fees	3,000	6,000	3,000			
	Total	123,000	126,000	123,000			
MONEY OUT	Production cost	89,100	103,950	118,800			
	Business permit fee	1,000					
	Artisan construction fees	2,500	5,000	2,500			
	Meeting cost (teas)	300	300	300			
	Total	92,900	109,250	121,600			
BALANCE		30,100	750	(1,400)			
START-UP CAPITAL	75,000						

- The groups cash situation is expected to deteriorate despite good business. By months 5 and 6 for Nthongoni and months 3 in Vaele group, the groups may not have money to meet some operating costs.
- Purchases are projected to increase while sales remain flat. Hence, money will be tied in un-sold stocks.
- Proper cash flow planning helps foresee such cash flow problems.

5.5. Summarize the session with the following points: 

- Money plans help foresee if you will experience cash shortages or surpluses in the future.
- If your Money Plan shows a cash shortage in a future period:
    - Manage your variable costs
    - Reduce purchases to minimize increasing stock levels

Step 6: 5 minutes

Learning Assessment
Training materials: Handout 1.4

Using Handout 2.0, ask participants to assess each of the three sessions above by ticking on the appropriate box as relevant.

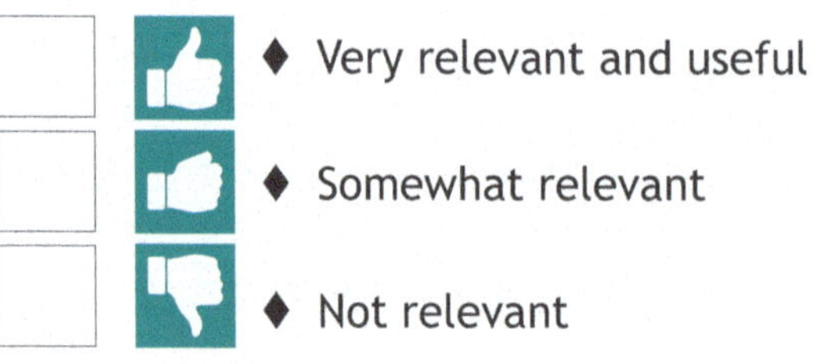

Bill of Quantities for Constructing SAFI Latrine

Hand out 3.1

Cost of a SAFI Latrine:

S/N	DESCRIPTION	UNIT	QUANTITY	RATE (KES)	AMOUNT (KES)
1	**Sub-structure**				
1.1	**Excavation of the pit latrine**				
1.1.1	Excavation to a depth between 5 feet-4 culverts depending on the soil structure (pit diameter of 1 m)	Feet	10	0	0
1.2	**Production of rings/culverts**				0
1.2.1	Ash (free)	Bags			0
1.2.2	Wire mesh (4 ft × 8 ft) - Light gauge	No	3	120	360
1.2.3	Portland cement (50 kg bag)	No	2	500	1,000
1.2.4	Sand (wheelbarrow)	No	4	120	480
1.2.5	Binding		1	100	100
	Sub-Total cost				**1,940**
1.3	**Precast slab**				0
1.3.1	Heavy gauge wire	Pieces	1	500	500
1.3.2	Red oxide	kg	1	150	150
1.3.3	Cement	Bags	1	500	500
1.3.4	Sand	Wheel barrows	2	120	240
1.3.5	Used oil	Liter	1	50	50
1.3.6	Labor (unskilled +skilled)	Man-days	1	1,500	1,500
	Sub-total cost				**2,940**
	Total sub-structure cost				**4,880**

Bill of Quantities for Constructing SAFI Latrine

Hand out 3.1 (cont'd)

S/N	DESCRIPTION	UNIT	QUANTITY	RATE (KES)	AMOUNT (KES)
2	Super-structure				
2.1	Super-structure walls				
2.1.1	River sand	Wheel barrows	5		
2.1.2	Ballast	Wheel barrows	3		
2.1.3	Bricks	Pieces	290		
2.1.4	Cement	Bags	3		
2.1.5	Timber	Feet	12		
2.1.6	Reinforcement wire (gage 12-twisted)	Pieces	1		
2.1.7	Wire mesh	Pieces	1		
2.1.8	Binding wire	kg	1		
2.1.9	Door	Pieces	1		
2.1.10	Hinges	Pair	1		
2.1.11	Locks	Pair	1		
	Sub-Total Cost				
2.2	Roofing				
2.2.1	Iron Sheets	Pieces	2		
2.2.2	Ordinary Nails-Assorted	kg	Quarter		
2.2.3	Roofing nails	kg	Quarter		
2.2.4	Vent pipe	Pieces	1		
2.2.5	Facila Board	Feet	12		
2.2.6	Labor (skilled)	Man-days	3		
	Sub-Total Cost				
	Total Superstructure (KES)				
	Total Cost/SAFI latrine (KES)				

My Money Plan
Handout 3.2

Weeks/Months	Description	1	2	3	4	5	6
MONEY IN							
MONEY OUT							
Total Money In							
Total Money Out							
Change in Money							
MONEY							

Participant Evaluation Form

Handout 3.3

Ask participants to provide feedback on each section by signaling with thumbs up, thumbs flat or thumbs down. Mark with ticks or numbers participants' feedback for each session in the boxes below.

Session	Very relevant and useful	Somewhat relevant	Not relevant or useful
Financial literacy			
Saving			
Borrowing			
Costing & Pricing			
Money Planning			

Timetable

Course: Business Skills Training for Sanitation Entrepreneurs
Module: Basic Financial Literacy

Session:	Venue:	Date:

Time	Activity	Facilitator
5 minutes	Ice breaker and introduction of session	
20 minutes	Financial literacy	
20 minutes	Saving	
20 minutes	Borrowing	
	Break	
30 minutes	Costing & Pricing	
20 minutes	Money Planning	
5 minutes	Learning Assessment	
	End	

Module 4

MANAGING YOUR
SANITATION ENTERPRISE

Business Skills Training for
Rural Sanitation
Entrepreneurs

CASH SALE

Session Plan

Session Tittle	Management and Leadership
Time	1 hour, 30 minutes
Objectives	At end of this session, participants should be able to:

◊ Differentiate between managing and leading

◊ Appreciate the importance of good leadership in a group enterprise

◊ Formulate an enterprise business proposition using the canvas model.

Contents

◊ Concepts of management and leadership

◊ Leadership as a critical attribute of an entrepreneur

◊ Business planning and modelling:
The BUSINESS CANVAS MODEL

Resources

◊ Local currency e.g KES 500 (in 50/- denomination notes)

◊ Sticky notes (at least three colors)

◊ Masking tape

◊ Three leaves of a flipchart

◊ Two pairs of scissors at least four manila papers of different colors

◊ A local sanitation entrepreneur

Methods

◊ Group exercise

◊ Question and answer discussion

◊ Practice sessions

CASH IN
OPPORTUNITIES
GROWTH
CASH OUT
COMPETITION
INNOVATION
EQUITY

Session Guide

Entrepreneur as a Leader and Manager

This session aims to develop the leadership capacities of participants to lead their small (startup) enterprises and grow their businesses. Participants will learn important leadership skills and traits of successful entrepreneurs in their locality and in Kenya.

1.1   Start by informing participants that a successful enterprise depends on its management – how competent the owner, leaders or member-owners are in leading and planning for the enterprise. This is the focus of this module, which builds on what they have learnt from start of training.

1.2   Give participants about 10 minutes to answer the questions below and then come back together to discuss their answers.

Group work guiding questions:

♦ Who is a leader and manager?

♦ List some of the characteristics of a leader or manager

♦ Mention one leader/manager in your community that you admire

1.3   Summarize the session by reinforcing on the following points:

♦ Reinforce their definitions using Handout 4.1.

♦ To be a successful entrepreneur, you must be a strong LEADER.

♦ You can delegate MANAGEMENT responsibilities, but not LEADERSHIP, especially of a small (start-up) business

Trainer's Notes

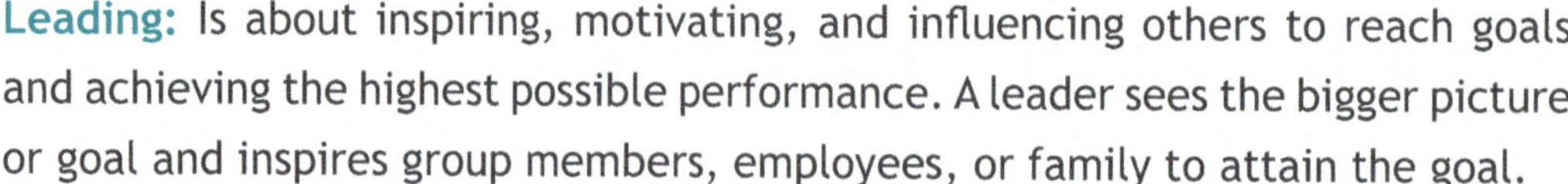

Leading: Is about inspiring, motivating, and influencing others to reach goals and achieving the highest possible performance. A leader sees the bigger picture or goal and inspires group members, employees, or family to attain the goal.

Managing: Involves organizing and overseeing daily operations and activities of your enterprise. It's the routine task of ensuring that the operations of the enterprise run smoothly.

Both are important, but clear, motivating leadership is critical for a successful enterprise.

The ability to motivate and inspire people within an enterprise directly relates to success in reaching goals. Leadership has three aspects; the leader, follower and the situation. The leader should be a person who can influence the followers in a given situation to realize the set objectives of the organization. The situation sets the context through which a given leadership style is exercised. Leadership requires the practice of integrity to enhance trust. Effective leaders provide opportunities for improvement and change, take risks, and focuses on the future.

Step 2: 30 minutes

Qualities of Good Leadership in a Group Enterprise

Training materials: Handout 4.1

2.1   Start this session by informing participants that for proper function of an enterprise, owned by family or group, essential tasks and responsibilities must be assigned to avoid any confusion or conflicts within the enterprise. In a group enterprise, leaders are selected based on consensus or elections to coordinate and manage its activities in their best interest. Leadership is therefore a critical function of the group — and requires persons with high integrity.

2.2   Inform participants that it's time for a game to explain the need for good leadership in group enterprise.

Paper Art - 15 minutes

Let participants go back into their two groups. From each group, pick one of the participant to remain as observers — looking at what the group is doing, who does what tasks, who gives directions, who talks most, who volunteers to do the cutting or who provides more ideas on what to do. And which group was most creative and did the work faster?

Provide the scissors and the manila paper to each group. Advise the group to cut the papers in any shapes they prefer, and prepare a collage. The group that prepares the best art will be rewarded.

The two observers will provide their observations on the group that had better leadership — the winner. Who among the group members showed more leadership characteristics? Who are they? What did they do?

2.3 Summarize the session by informing participants some of the important qualities of a good leader within a group:

♦ Democratic and allows all members to participate in group activities

♦ Self-disciplined — leads by example

♦ Patient and acts in the best interest of the group enterprise

♦ Honest in dealing with group enterprise activities

♦ Demonstrates clarity in his business vision

♦ Effectively manages internal conflicts, aligning all to one goal

Exercise:

Is there a local entrepreneur who exhibits similar business leadership and management characteristics in your local area?

Is there a person within their group who exhibits some of these characteristics?

Note: The trainer has an option of inviting a successful local entrepreneur to share his/ her experience. The trainer can also choose to read Dr. James Mwangi's story on Hand out 4.1 to highlight and summarize the attributes of leaders discussed in the exercise above.

Step 3: 30 minutes

Developing your Business Plan: The Business Canvas Model

Handout 4.2 : Business Canvas Model

3.1 Inform participants that this last session brings together all the elements they have learnt from Module 1 to up to this stage of Module 4. They will learn how to develop a one-page business plan for their enterprise in a very simple and fun-filled way.

3.2 In this session participants will learn to prepare a simple business plan using a tool known as BUSINESS CANVAS Model. The business canvas offers a simple structure of developing a business plan by just answering a set of nine-key questions.

Refer to Handout 4.2 on how to fill the Business Canvas Model.

Refer to Module 2 on answers to these questions:

1. Who are your customers?
2. What's compelling about the product or service you offer? Why do customers buy, use?
3. How are you delivering your product or service to customers?
4. How will you keep maintain a good relationship with customers, so that they keep coming back or talking about your offering to others?

Refer to Module 3 on answers to these questions:

5. How do you make money from the product or service you offer?
6. What main activities do you carry out, and which costs money, in order to deliver your product or service to customer?
7. What important resources, in terms of assets, expertise and others do you need to better serve the customer?
8. Which are your major costs that impact on the revenue you generate? Look at activities and resources above

Refer to Module 1 on Partnerships for Answers

9. Who are the Key business partners — suppliers of products or services and other organisations who are key for your business to succeed?

3.3   For this session, participants should refer to the blank Business Canvas template in their handbooks.

Activity

- Then using a flip chart, draw the nine cells of the Canvas and place on the wall, using Handout 4.3 as a guide.

- Using one of participants or group enterprise, have participants slowly fill in the cells from the TOP RIGHT to the LEFT using sticky notes of different colors. The idea is to link each customer segment and their value proposition using a same color. See sample below:

3.4   Summarize this session with emphasis on importance of business planning:

- ◆ A business that never plans is planning to fail.

- ◆ Successful entrepreneurs know their customer needs, operating cost and the right pricing level to remain profitable.

- ◆ Always look back and review your business plan and strategy as the market evolves.

Step 4: 5 minutes

Learning Assessment
Training materials: Handout 4.8

Using Handout 4.3, ask participants to assess each of the three sessions above by ticking on the appropriate box as relevant.

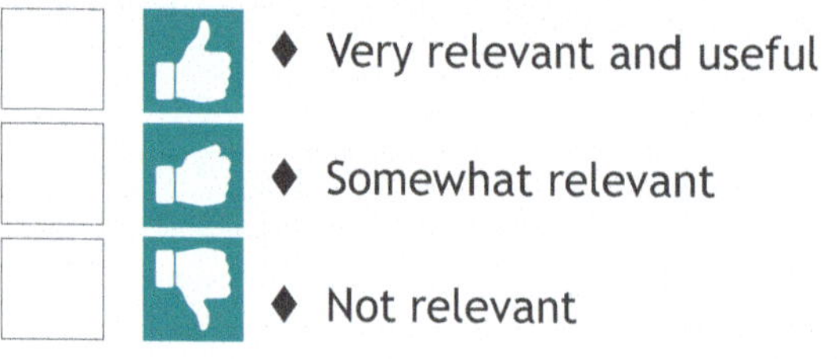

Handout 4.1

Leadership Case Study: Dr. James Mwangi

Dr. James Mwangi hails from the beautiful highlands of the Aberdares in Murang'a county. His father was a Mau Mau fighter who died in the struggle, leaving behind a young widow with seven children.

Growing up in Kangema village was challenging. Just like any other low-income family, money was scarce and the family teamed up to supplement their income by engaging in small businesses. There was no time for childish games — everyone had to pitch in to keep the home fires burning. Mwangi, like the rest of his siblings, had to do his share of chores — tending to livestock, making charcoal, selling fruits and other produce for small margins. Dr. Mwangi specifically sold charcoal and fruit and tended livestock to help support his mother and siblings.

While this may have been humbling for the young Mwangi, he was nevertheless absorbing invaluable business lessons that would stand him in good stead for his future. He was learning, without consciously doing so, the basics of business — what people needed, what they were prepared to pay, how to add value to products, the basics of customer care, negotiation skills, how to make a sale and turn a profit. With no role models to emulate, he and his family were, in effect, discovering the basics of business all by themselves, based on observation of what worked and what did not. This motivated Mwangi to study commerce at the university.

Dr Mwangi is one of the founders and CEO of Equity Group which owns Equity Bank, one of the largest banks in East Africa. Through his leadership, Equity bank was the first bank to successfully offer micro-financing products to low-income earners and small enterprises — including sanitation businesses — in Kenya.

Dr. James Mwangi is today one of the country's most influential persons in entrepreneurship and one of the most successful businessmen in Africa.

Business Model Canvas

	Enterprise:	Trainer:	Date:

Key Partners	Key Activities	Value Propositions	Customer Relationships	Customer Segments
Who are our Key Partners? Who are our key suppliers? Which Key Resources are we acquiring from partners? Which Key Activities do partners perform?	What Key Activities do our Value Propositions require? Our Distribution Channels? Customer Relationships? Revenue streams? **Key Resources** What Key Resources do our Value Propositions require? Our Distribution Channels? Customer Relationships Revenue Streams? TYPES OF RESOURCES: Physical, Intellectual (brand patents, copyrights, data), Human, Financial	What value do we deliver to the customer? Which one of our customer's problems are we helping to solve? What bundles of products and services are we offering to each Customer Segment? Which customer needs are we satisfying?	What type of relationship does each of our Customer Segments expect us to establish and maintain with them? Which ones have we established? How are they integrated with the rest of our business model? How costly are they? **Channels** Through which Channels do our Customer Segments want to be reached? How are we reaching them now? How are our Channels integrated? Which ones work best? Which ones are most cost-efficient? How are we integrating them with customer routines?	For whom are we creating value? Who are our most important customers? Is our customer base a Mass Market, Niche Market, Segmented, Diversified, Multi-sided Platform

Cost Structure	Revenue Structure
What are the most important costs inherent in our business model? Which Key Resources are most expensive? Which Key Activities are most expensive?	For what value are our customers really willing to pay? For what do they currently pay? How are they currently paying? How would they prefer to pay? How much does each Revenue Stream contribute to overall revenues?

Participant Evaluation Form

Ask participants to provide feedback on each section by signaling with thumbs up, thumbs flat or thumbs down. Mark with ticks or numbers participants' feedback for each session in the boxes below.

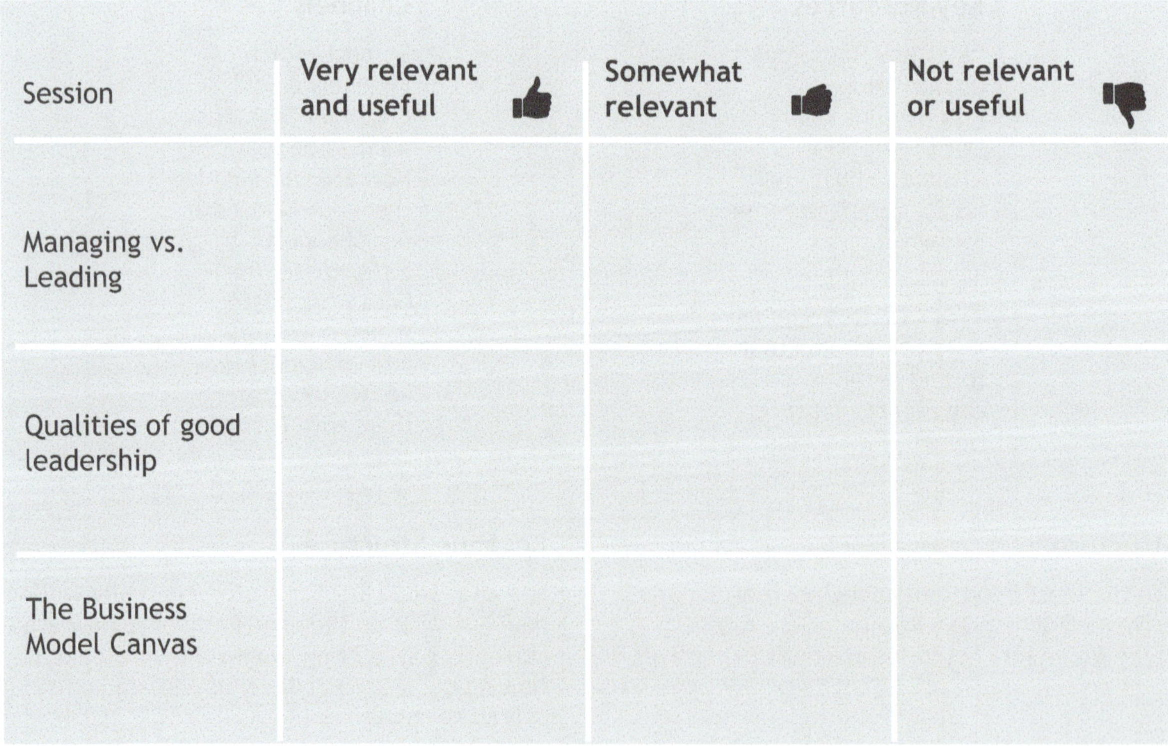

Session	Very relevant and useful 👍	Somewhat relevant 👊	Not relevant or useful 👎
Managing vs. Leading			
Qualities of good leadership			
The Business Model Canvas			

Handout 4.4

Timetable

Course:	Business Skills Training for Sanitation Entrepreneurs
Module:	Management and Business Leadership

Session:	Venue:	Date:

Time	Activity	Facilitator
5 minutes	**Ice breaker and session introduction**	
15 minutes	Managing vs. Leading	
20 minutes	Characteristics of successful business leaders	
	Break	
30 minutes	Creating a vision statement	
40 minutes	The business model canvas	
5 minutes	Learning assessment	
	End	

REFERENCES

Dumpert, J., & Perez, E. (2015). Going beyond mason training: enabling, facilitating, and engaging rural sanitation markets for the base of the pyramid. *Waterlines, 34*(3), 210–226. http://www.jstor.org/stable/24688132

DSW (Deutsche Stiftung Weltbevoelkerung). (2009). *Entrepreneurship Development Training Manual* (1st ed., Issue 1).

Joseph, V. H. C. F. P. (HCFP). (2003). *Training Manual on Self-Help Groups for Development.* (Issue 2nd).

Kibet. J, Ochieno. C. (2019). *Financial Literacy for Smallholder Farmers.* Kenya National Library Services. Nairobi, Kenya. ISSBN: 978-9966-825-10-0

Meijerink, R. (1994). *Simple Bookkeeping and Business Management Skills* (second edition, Issue 1). http://agris.fao.org/agris-search/search. do?recordID=XF2006421419

Mullanji Ana; Topalli Irena. (2017). *Training Module on Entrepreneurship* (first edition, Issue 1).

Nabembezi Dennis, N. H. (2019). *A Handbook for Sanitation Managers and Private Sector Players. Journal of Chemical Information and Modeling* (Vol. 53, Issue 9). https://doi.org/10.1017/CBO9781107415324.004

Water and Sanitation Programme (WSP), Laos Government, P. (n.d.). *Business Skills Training for Sanitation Entrepreneurs: Facilitator's Guide.*

Water and Sanitation Programme (WSP), Laos Government, PSI. (n.d.). *Latrine Sales Agent Trainee Handbook.*

NOTES

IWA Publishing's authorised EU representative for General Product
Safety Regulations is Diane D'Arras, 15 rue Duret, 75116 Paris,
France, e-mail: safety@iwap.co.uk.

Printed and bound by CPI Group (UK) Ltd, Croydon, CR0 4YY
12/05/2026
02108887-0002